GOOD SCIENTIST

Abhijit Naskar is the twenty-first century Neuroscientist whose contributions in Cognitive and Behavioral Neuroscience have helped the world tackle the issues of mental illness, prejudice, hate, extremism, discrimination and segregation more effectively. As an untiring advocate of mental health and universal acceptance, he became a beloved best-selling author all over the world with his very first book "The Art of Neuroscience in Everything". With his pioneering ventures into the Neuropsychology of beliefs and biases, he has hugely contributed in the eradication of religious and cultural differences in our world, for which he is popularly hailed as the humanitarian scientist, who takes the human civilization in the path of sweet general harmony.

GOOD SCIENTIST
When Science and Service Combine

ABHIJIT NASKAR

Good Scientist: When Science and Service Combine

Copyright © 2020 Abhijit Naskar

This is a work of non-fiction

All rights reserved. No part of this publication may be reproduced, distributed, or transmitted in any form or by any means, including photocopying, recording, or other electronic or mechanical methods, without the prior written permission of the author, except in the case of brief quotations embodied in critical reviews and certain other noncommercial uses permitted by copyright law.

An Amazon Publishing Company, 1st Edition, 2020

Printed in the United States of America

ISBN: 9798557808835

Also by Abhijit Naskar

The Art of Neuroscience in Everything
Your Own Neuron: A Tour of Your Psychic Brain
The God Parasite: Revelation of Neuroscience
The Spirituality Engine
Love Sutra: The Neuroscientific Manual of Love
Homo: A Brief History of Consciousness
Neurosutra: The Abhijit Naskar Collection
Autobiography of God: Biopsy of A Cognitive Reality
Biopsy of Religions: Neuroanalysis towards Universal Tolerance
Prescription: Treating India's Soul
What is Mind?
In Search of Divinity: Journey to The Kingdom of Conscience
Love, God & Neurons: Memoir of a scientist who found himself by getting lost
The Islamophobic Civilization: Voyage of Acceptance
Neurons of Jesus: Mind of A Teacher, Spouse & Thinker
Neurons, Oxygen & Nanak
The Education Decree
Principia Humanitas
The Krishna Cancer
Rowdy Buddha: The First Sapiens
We Are All Black: A Treatise on Racism
The Bengal Tigress: A Treatise on Gender Equality
Either Civilized or Phobic: A Treatise on Homosexuality
Wise Mating: A Treatise on Monogamy
Illusion of Religion: A Treatise on Religious Fundamentalism
The Film Testament
Human Making is Our Mission: A Treatise on Parenting
I Am The Thread: My Mission
7 Billion Gods: Humans Above All
Lord is My Sheep: Gospel of Human
Morality Absolute
A Push in Perception
Let The Poor Be Your God
Conscience over Nonsense
Saint of The Sapiens
Time to Save Medicine
Fabric of Humanity

Build Bridges not Walls: In the name of Americana
The Constitution of The United Peoples of Earth
Lives to Serve Before I Sleep
When Humans Unite: Making A World Without Borders
All For Acceptance
Monk Meets World
Mission Reality
Citizens of Peace: Beyond The Savagery of Sovereignty
Operation Justice: To Make A Society That Needs No Law
See No Gender
The Gospel of Technology
Every Generation Needs Caretakers: The Gospel of Patriotism
Aşkanjali: The Sufi Sermon
Mad About Humans: World Maker's Almanac
Revolution Indomable
When Call The People: My World My Responsibility
No Foreigner Only Family
Hurricane Humans: Give me accountability, I'll give you peace
Ain't Enough to Look Human
Servitude is Sanctitude
Time To End Democracy: The Meritocratic Manifesto
I Vicdansaadet Speaking: No Rest Till The World is Lifted
Boldly Comes Justice: Sentient not Silent

DEDICATION

*This book is dedicated to
Vilayanur Subramanian Ramachandran,
I am merely dust under your feet.*

CONTENTS

1. When Comes The End 1
2. Why We Seek Approval 5
3. Implications of Attention Seeking 11
4. Beyond Conformity and Non-Conformity 17
5. Gospel of Science 23
6. Sonnet of Paths 31
7. Kindness is No Philosophy 35
8. Walk Past The Delusion of Status 41
9. Sonnet of Behavior 45
10. Gospel of Worship 49
11. When Realization Calls 55
12. Gospel of Good Government 59
13. Gospel of Good Citizenry 65
14. Observation over Cynicism 71
15. Goodwill Gospel 75
16. When Calls The Future 81
17. We Are All Idiots 85
18. Arise, Generation Assimilation 89
BIBLIOGRAPHY 95

1. When Comes The End

ABHIJIT NASKAR

Let's start with the end shall we - what are you going to leave for your children? What is going to be the most precious thing that you are going to hand over to your children? What is going to be your keepsake for your children?

If you want to leave something for your children, leave a better world, not heaps of money, because at the rate our ancestors screwed up this world and at the rate we are sustaining their stupidity in our pursuit of limitless productivity, soon all the money in the world will not be enough to save our children from imminent disaster. And here the question is not if, but when - unless you actually wake up the neurological marvel called accountability inside your heart and act upon that accountability in the path of genuine humanity.

Remember, productivity is not necessarily a good thing. Productivity with no moderation - with no conscience - with no vision of its implications in the society, doesn't ensure progress, it only takes us closer to an absolute collapse of civilization. And that's precisely the direction in which we are headed.

Whether we head towards the collapse of civilization on a golden chariot or on a run-down horse cart, it doesn't change the destination. And with all our scientific marvel, especially technological ones we have built ourselves a golden chariot on which we are proudly headed towards disaster - all while we are absolutely immersed in the tell-tale pleasures of progress.

If you know a lot of science, you are a database, but if you can use the science for the benefit of society, then only you are a scientist. One is a machine, the other is a human - so the question is, what are you? To make it even simpler - let me put it this way - there is no such thing as a good scientist and a bad scientist - there is only scientist and machine.

Even the most powerful machine is worthless compared to a true scientist, for no matter how powerful a machine is, it's still a machine, it's a dead pile of data, nothing more - a scientist on the other hand is a vessel of infinite possibilities - possibilities that are beyond computation - quite literally speaking.

2. Why We Seek Approval

ABHIJIT NASKAR

Human potential is incomputable. We the experts in cognitive and behavioral sciences can predict human behavior but not human potential. What this means is that, though we can tell how a person is likely to feel, think and behave in a certain situation, we still cannot tell what a person is capable of. Hence the possibilities that a person holds in their neurons are immeasurable.

We can predict human behavior only because people have a tendency to behave like everybody else. Even when they try to act different, they do so to attract attention, either consciously or subconsciously. You see, there is a difference between acting different and being unique.

The more people act different, the more they act the same.

People don't act different because they are different, they act different to look different, because to look different is to look attractive. Hence it's all about attention craving, not uniqueness, not truthfulness, not being who they are. But again, this very trait of craving for attention from one's society is also a common part of human nature.

By seeking attention from our society, that is, by seeking approval from our society we subconsciously desire to be accepted by the society - why you ask - imagine you are stranded in a jungle alone - how long do you think you'll survive - now imagine you are stranded in a jungle but you are not alone, you have an entire community of humans with you - now the chances of survival for not just you, but your entire community increase exponentially - and this is exactly how our ancestors in the jungle survived, and though we have built our own civilization now separate from wildlife, the tenets of our old days remain pretty dominant in our neurobiology.

Therefore, seeking approval from the society is nothing abnormal - on the contrary, it's the very norm of the human society. However, there's a twist in this story. By seeking approval from others and by behaving like everyone else we may manage to survive for a short while, but we won't be making any progress whatsoever, for progress happens on the shoulders of those who are original and actually, genuinely do not care much whether society approves of their originality or not - they do not care much whether they are mocked or applauded - they do not go against society mark you, rather they just do what defines who they are, which is usually termed by the society as an act of rebellion, because they do not do what the masses do.

3. Implications of Attention Seeking

Originality breeds progress and collective commitment of the people to the achievements of that progress sustains those achievements, which means that if the people of a society do not feel strongly about the achievements of their time, no originality can keep those achievements alive. The more a people feel strongly about something, the more they want to express those strong feelings with others, and the more they share with others, the stronger those feelings become, both in their individual psyche as well as in the collective psyche of the society.

Let me give you an example. Social media has gained such popularity in such a short period of time because it has given people the easiest way to express themselves. But here's the catch. Freedom of expression is innately intertwined with the desire for social approval or appreciation or attention. You only feel like expressing something special where there is an opportunity of being noticed by others.

Let me explain. You do not dress up in nice clothes to snuggle up on your couch and watch a movie, yet when you go out with friends or with a date on a dinner or to go to the theatre you dress up nicely. Why do you think this is –

it's because when you are at home, all by yourself, you do not have the need to grab attention of others, hence there is no need for you to "express yourself specially" - at home you just are who you are. But when you go out and there are people around, whether they are friends or your date or just plain strangers, your mind automatically kicks into the primitive approval/attention seeking mode and coaxes you to look appealing to others, quite without your knowledge as to why.

This attention seeking is not wrong or right - it's just part of human nature, but it can turn lethal for the individual human as well as the society, when practiced to the extreme without any awareness and moderation whatsoever. For example, the very existence of social media is predicated on humankind's primitive drive of attention seeking. And when they successfully monetize your attention, they end up with billions of dollars and you end up with a screwed up mental state. And if we don't do anything about it now, the next generation will be a generation of mentally unstable glass creatures.

So how do we solve this appalling problem! The first step is to recognize the problem and make the platforms - not ask mark you, but make them accountable - legally that is. To achieve that I propose the following solution.

I urge all responsible citizens across the world to demand the parliament of your country to ban the use of any social media platform that doesn't have a health hazard warning on their welcome screen stating "excessive use of this platform can cause severe mental health problems".

Then the next step would be to legislate nation-wide social media awareness campaign for all citizens of all ages. These two can take place side by side even - the order is irrelevant, as long as they are in action.

The desire for expression is intrinsic to our existence, that is, to human existence, and nothing can take that away, however, that desire should be driven primarily by the pleasure of expression itself and not by the craving for attention or recognition - like the genius painter who paints not for applause but because the strokes of paint reflect their very soul - or like the genius writer who writes not to gain recognition but because the words reflect their very being. Let expression be the reason for expression, not attention.

You can have one or two persons in your life to whom you can express without worrying about whether you are expressing because you want to express or because you want their attention, for in these very personal circumstances the line is nonexistent - but as for the rest of the society is concerned, be very aware not to express with the expectation of attention.

There will be expectation of course, for as I said earlier, it's a part of the human psyche, however, all I am asking is that you be aware of those expectations and gently nourish your mind with the power of will to expect less and less - and the more you learn to be aware of the expectations the more you'll become better at reducing them. It's like keeping your biases in check - the more you are aware of your biases the more you can choose whether or not be driven by them.

4. Beyond Conformity and
   Non-Conformity

Neither conformity nor misbehavior is the key to a civilized society, awareness and accountability are. The problem is, people think that the opposite of conformity is recklessness and misbehavior - but it's not - the opposite of conformity is actual, genuine originality. Certain norms of the society must be accepted, because if you reject society on all aspects, then why should society accept you on any aspect - and if you say you do not need to be a part of society, if you say that you prefer being on your own, then you have no place being in a civilized world to begin with - you are better off living in a jungle somewhere alongside your fellow animals - for if a person narcissistically cuts off all ties with society, they cut off all ties with humanity.

Human society is far from perfect, it is far from civilized, but it's our responsibility to turn that less civilized society into a more civilized one - and in that endeavor a certain amount of solitude is indeed necessary to be aware of your abilities, but once you have realized your abilities, it's time for you to bring those abilities out and caste them into the society so that the so-called civilized world can become actually civilized.

Norms are essential to sustain the fabric of society, but no norm should be accepted without scrutiny, the moment we do is the moment we fall, not just as individuals, but as the collective. Norms are necessary, so is original thinking - neither one is expendable.

Norms without original thinking brings societal progress to a halt, original thinking without norms makes the society unstable. So do not blindly accept whatever is the norm, and at the same time, do not blindly chase after anything that is different. Scrutinize everything - scrutinize mark you, not criticize. Criticism is the weapon of the shallow, scrutiny is tool of the wise.

First you must learn to distinguish between a difference in opinion and an act of injustice. Someone may disagree with you on certain matters, but that doesn't mean they are being unjust or inhuman. Let me give you an example. I have a dear friend who is a prominent mathematician of the world, and we often disagree when we start talking about the human mind, but our friendship remains strong as ever.

So stand up to injustice and inhumanity wherever you witness them but do not criticize or hate people just because they disagree with you. I have said this before and I say this again - better lose an argument than a friend. Argumentation doesn't breed solution, it only breeds more chaos. If you want to find solution, then observe, scrutinize and discuss.

Discussion can solve problems that guns and pills cannot.

However, a discussion has to be an actual discussion to be able to solve a problem, and not a debate in disguise. In most cases, when people engage in discussion they do so not with the desire to actually discuss a matter, but to prove the significance of the opinions that they've already concocted prior to the discussion. So, discuss, don't debate.

ABHIJIT NASKAR

5. Gospel of Science

Science is founded upon the very faculty of scrutiny, even though quite like the lay people, many scientists confuse scrutiny with criticism - after all, scientists are humans as well, and as such, even they are not infallible. In the insane pursuit of seeing their name below the title of a research paper, scientists have forgotten the very purpose of science - that is - to serve the society. Science that doesn't serve the society is not science but con-science.

There are two kinds of scientist, those who are in science merely to gain recognition by publishing paper after paper and then there are those who are in science with the genuine desire to help the people - the former are mere hypocrites and absolutely unworthy of the title scientist, whereas the latter are not only scientists but actual living, breathing human beings with an indomitable sense of responsibility towards the society.

In science it's not the research that counts, but how it helps the people.

Any field that forgets the people is a field of the dead, it has no business being in a civilized society. Science without responsibility is science of the savages.

Here some intellectuals may most profoundly argue – as if savages are capable of science - and my answer to them is, why not - in fact, science has its origin in our savage days - our savage ancestors devised tools and weapons out of wood and rocks - they even learnt to tame and create fire - what do you think these were! These were the earliest scientific achievements of humankind, which followed by many more eventually placed humankind above all life on earth.

In fact, modern humans with guns and nukes are no different from cavemen with hand axes. The point is, when we first started exploring the domain of science quite unknowingly, we had no idea in the first place as to what we were doing. But we no longer are reckless savages anymore and the power of science that we wield in our hands today if practiced without accountability would do more harm to the society than good.

Hence we cannot place the science of today at the same level as the science of yesterday, for as

I have said before, everything in the human world must evolve with time, if it doesn't it either gets destroyed or destroys the world.

Religion that doesn't evolve destroys the world in the name of faith and science that doesn't evolve destroys the world in the name of advancement.

It's better to have no science than to have a science without accountability. A scientist without accountability is like a kid with a candle - they may feel the power of the candle but if they don't realize its real life implications, they'd do more harm than good. Science is like a candle, it can light up a room or it can burn down a house - in the hands of the responsible scientist it lights up the world but in the hands of the empty scientist it'll burn the whole world to ashes.

Therefore, mere advancement of science won't do any good to the society unless that advancement is guided by the responsible hands of good scientists - scientists in the line of service - scientists in the line of sacrifice - scientists in the line of devotion, not merely to the pursuit of knowledge and abolition of ignorance, but more importantly to the pursuit of knowledge and abolition of ignorance with the express purpose of benefitting the world.

The problem with various fields of understanding today is apathy. Take medical science or medicine for example. Often mentors in medicine teach their students that attachment with the patient is unprofessional in the practice of medicine. It is one of the ghastliest errors of human intellect. Attachment is what makes us human, so when you say attachment is bad, you are saying life is bad. We need not less attachment, but more attachment in our world - attachment to not just our immediate family, but to our neighbors - to our friends - to the neighbors across the border - to the neighbors from the other side of the world.

**Till the whole world becomes our family,
no family can live in peace.**

By wielding the power of science in hand, humankind have become the masters of an entire planet - but the question that we should be asking is - is it really the humankind that have become the masters of the planet, or is it only the privileged class with the access to the scientific wonders who are the real masters! Do not rush for an answer - just think over the question for a while.

Science if put to real use for the benefit of humanity could eliminate suffering in a matter of months, yet suffering persists - you know why, because the science of today is driven by capitalism not curiosity. Science driven by capitalism creates a divided planet - where one section has more than they can digest and another has not even the essentials to keep body and soul together.

Science is power but what good is that power if it doesn't help those who really need help - what good is that power if it only aims to increase the comfort of those who are already living in comfort! Some may say, even though science may not aim to help the downtrodden directly, in the long run it benefits all - and though the statement is factually correct, it still is the coldest statement that any so-called human could ever make. Elevating all of humanity ought to be the highest priority of science, not the lowest.

What kind of cold, savage, nincompoop says that the only way science can help the helpless is through collateral benefit - which means, let's serve the privileged first and earn more revenue, and somewhere along the way the helpless will be benefitted somehow. This is not science, this is ignorance and bestiality at their prime.

Alleviation of human suffering should be the direct outcome of science, not a collateral byproduct.

And this is precisely what I have been doing with my work - by putting the science of human nature to use in the course of raising a just and unified society.

Ten average scientists with a strong sense of humanity can do more good to the world than a hundred brilliant scientists with weak humanity. Science is power - and the more power you have, the more responsible you must become, if not, then that power will do more harm than good. Science is a tool, religion is a tool, philosophy is a tool, politics is a tool, none of them has any conscience of its own, it's the people using those tools who determine their implications in the society - who determine the nature of those tools in the society.

6. Sonnet of Paths

Sonnet of Paths

Science means nothing,
Unless we use it to lift the society.
Philosophy means nothing,
Unless it empowers humanity.
Religion means nothing,
Unless it advocates for inclusion.
Technology means nothing,
Unless it aids in collective ascension.
Tastes are plenty in our world,
So are the paths that humans take.
But if those paths hold no humanity,
Fabric of civilization will soon break.
Placing on humanity our prime attention,
Together we'll attain true emancipation.

7. Kindness is No Philosophy

Society is us - and I mean it literally - society is not some magical construct that exists devoid of humans - society is the collective expression of the individual humans - so when the individual humans become responsible of their actions in society, not in the name of civility mark you, but when they actually feel accountable of their actions, the society automatically turns into a civilized one. In such a society, ideas like altruism, existentialism, humanism, socialism and all these philosophical and political ideologies would exist as part of history not as part of the present.

Let me elaborate further with the idea of altruism. You see - altruism is a myth - it's an intellectual invention that tries to philosophize kindness. It's this kindness that defines a true human, yet the concept of altruism attempts to alienate the act of kindness by placing it in an exclusive philosophical circle separate from everyday, ordinary humanity, and in doing so it fails to recognize the very characteristic that makes humanity human, just like the fundamentalist circle fails to recognize the core essence of religion by trying to imprison it into exclusive scriptures.

Here's the simple fact of the matter. Kindness is no philosophy, it's no theological concept, it's the plain, ordinary, everyday responsibility that a human holds towards the society for being human. In short, kindness is humanity. And it's in this kindness that lies the seed of worship - it's in this kindness that lies the seed of holiness.

It's this simple - when you feel accountable - and I mean really, actually, genuinely accountable for your society, the right action appears in your mind on its own - then you no longer feel the urge to disprove others to prove yourself - then you no longer you feel the urge to diminish others to elevate yourself. When you diminish others to elevate yourself, you only diminish yourself further.

You cannot rise above darkness if all you try to do is shove others into darkness.

Light doesn't hate - light doesn't discriminate - light doesn't behave proud - light just acts as light and all darkness withers away in front of its sanctifying rays. Light doesn't say 'o look at me, I am so bright' - all it does is shine. And that's all you got to do if you genuinely want to eliminate inhumanity from the society - be the embodiment of actual human behavior - without pride - without hate - without prejudice - without discrimination - just being a human being – and all inhumanity will fade away on their own. It's the law of nature - savageries exist in the absence of accountability. Be accountable and all savageries would disappear.

8. Walk Past The Delusion of Status

You could be a scientist, you could be a teacher, you could be a preacher, you could be a janitor, you could be a construction worker, you could be a waitress, you could be absolutely anything in the world, your profession has no bearing over your capacity to do good to the society - feel accountable, think accountable, act accountable - that's the golden principle of societal health, sanity and upliftment. It's your accountability that makes you a human, not your profession. Let me put it into perspective with a simple statement. An accountable waitress is more capable of running a nation than a bigoted scholar.

The problem with scholarship is that, in the insane pursuit of knowledge, scholars lose touch with plain, ordinary, everyday living, and a scholar who loses touch with everyday life, loses touch with reality and one who loses touch with reality is no longer a part of society and one who is not a part of society is not a scholar to begin with, for the term scholar is a revered word that is used to refer to those individuals whose insight has a beneficial effect on society.

No matter what you are - no matter what your livelihood is - no matter what your status in society is - above all that you are a human being - a responsible, conscientious, alive human being - and being an alive human being means taking responsibility - it means being accountable - it means consciously being a thread on the fabric of society.

It is this simple - you are either an active part of society or not even a human. And that very activeness - that very eagerness to be involved in the affairs of the society - that very indomitable desire to take the problems of your society on your own shoulders is the real worship of the civilized society - it's the real holiness of an advanced species – in short, in this purposefulness lies the true beauty of humanity.

9. Sonnet of Behavior

Sonnet of Behavior

The beauty that you see with your eyes,
Is but an illusive sign of fertility.
The beauty that you see with your mind,
Is a sign of life, truth and eternity.
The peace that you seek in possessions,
Is but a mirage most rotten.
The peace that is dormant in your heart,
Will make this world awakened.
The order that you seek in law,
Is but a sign of disorder and inhumanity.
The order that the world truly needs,
Is born of your own accountability.
Chasing illusions breeds only insecurity.
Pursue meaning and there'll be serenity.

10. Gospel of Worship

Kindness is beauty – kindness is humanity – kindness is sanctity. It's the kindness of the people that makes a place holy, not the churches and temples. Let me tell you story. Once upon a time there was a wealthy king. He had a huge garden and he employed two gardeners to take care of it. One of them was absolutely useless - he would lazily spend all his time without doing a single work around the garden, but whenever his master visited, he would simply usher him with praise. The other gardener was a humble fellow and rather a quiet one. He didn't talk much, but he worked the whole day taking care of the garden and grew all sorts of beautiful flowers and fruits. Now who do you think would be most beloved to the king! Many people call this king God, I call it nature. And one who takes care of the children of the king is more beloved to the king than the one who does nothing but talks highly of the king.

To light one candle in the cottage of a poor villager is a thousand times holier than lighting a thousand candles in the church.

It is our world and taking care of it is our existential duty - call it humanity, call it holiness, call it kindness - the term is irrelevant, what's important is the act and the act alone. Remember, there is no road to holiness, for holiness is the road - either you walk on it or you don't, that is, either you act as a being of conscience and character in your everyday ordinary life or you don't.

Most people confuse divinity with magic, but here is the reality of the matter - there is no magical or extraterrestrial divinity involved in the affairs of the human world - the divinity that can make any difference in our world is the divinity that we hold in our heart - the divinity which is commonly known as humanity. You may attend a thousand churches and yet die as a divine vacuum, but one who does everything in their power to help others with or without any special inclination for the church, is the very embodiment of divinity.

Churches don't hold divinity, human hearts do.

Recognize that divinity, realize that divinity and bring out that divinity, for it's the only divinity that can rescue our world from the clutches of prejudice, discrimination and suffering. It is irrelevant whether you call it divinity or holiness or compassion or kindness or anything else - the only thing that matters is that you realize it and act upon that realization.

11. When Realization Calls

Realization is the mother of change - if you can realize, you can make it materialize, not by magic, but by tangible human action - where there is action, there is change - as a matter of fact, action is life, inaction is death. Life always acts - life always evolves - the moment it stops evolving is the moment it turns dead - the moment life stops acting, is the moment it stops being life.

So wake up my friend - wake up from death - wake up into life and act as a living being of conscience and character ought to act - how is a living being of conscience and character ought to act you ask - you have to find that out for yourself - the individual must find their own way, the same holds true for a nation - the same holds true for the world. And when the individual learns to find their own way, the world will learn it too, for the world starts with the individual.

You are not going to find the way in books, old or new, including those by me - books can clear a few things, that's all. Books are just a tool, quite like science itself, how you use them it's up to you. You can do all sorts of heinous deeds and use passages from books to justify them, or you can use them to strengthen your stance as a human - that is, a human with reason and warmth.

You don't have to accept everything that's written in a book, take the parts that are compatible with your life and time and put them to practice. Many people often boast about the huge number of books they read, but remember, it's not about how many books you read, it's about how you use those books in practice.

A whole life is not enough to learn everything - in fact, the moment we stop learning is the moment we die. The more we learn, the more we feel responsible - the more we are responsible, the more we want to learn. Learning fosters accountability and accountability fosters learning. And this applies to every single person on earth - every single human that is. Societal issues thrive on the indifference of the people – those issues disappear with rising accountability. In fact, society, human society that is, goes hand in hand with accountability.

12. Gospel of Good Government

A truly accountable government can end economic disparities within months, but no politician has either the guts or the desire to do so, for the very survival of a government is predicated on sustaining the issues of the society, not solving them. How can the government end economic disparities you ask - and the answer is simple - increase tax rate in proportion to income of every single citizen, no matter who they, no matter what status they hold in society.

In fact, all charities would disappear from earth once the governments start taxing the rich 90% of their income and investing that revenue in public essentials – such as, groceries, housing, healthcare and education. And this is no socialism - it's plain common sense.

Till luxury becomes a thing of the past, equality will remain a thing of the future.

And it is no grand task to accomplish either - for if the government really wants, it can make it happen immediately, that is, eliminate inequality, but it won't, instead it'll give you all sorts of poppycock excuse, such as 'legislation takes time' - or 'Rome wasn't built in a day' -

and all that - the real reason is, politicians in general never really serve the people, they serve special interest groups - they serve those very rich that the government ought to tax. Hence, the rich keep getting richer and the poor poorer, all because the government of today is built without a backbone - it's built without a marrow - its veins do not carry principles and dignity - all it's filled with is psychological sewage water.

Billionaires are the new monarchs of the world.

In the old days affairs of the state were dictated by kings and queens, today they are dictated by billionaires. Governments used to lick the boots of kings and queens, now they are the spineless yesmen to billionaires. Hence, taxing those billionaires is out of the question.

In a truly civilized society there wouldn't be any billionaire, nor will there be any homeless, for all the revenue generated through taxing the rich would be distributed among the people through welfare initiatives. It's this simple, the more a person earns, the more they ought to be legally bound to give back to the society – any government who has the guts to follow this simple principle can alleviate societal suffering in a matter of months.

But again, once a government successfully legislates exponentially higher tax rate for the rich, it would inadvertently give the politicians authority over all that money, so unless you choose those politicians wisely, it won't matter whatsoever whether they tax the rich or not, for so long as meritless, backboneless, virtueless politicians are in power, income tax revenue will only serve the interests of those politicians, not that of the people.

However, if the authorities in the governments were beings of merit and character in the first place they won't need to be reminded by me or any other scientist or scholar what the duties of a government are, for they themselves would feel the genuine urge to do everything in their power to benefit the society and end all its disparities.

Therefore you see, it all comes down to this - if the citizens are careful and place nothing but the very best in the position to take care of the society, then all the issues that haunt our society would disappear on their own. So it is all in your hands, not in the hands of the politicians, not in the hands of the billionaires, not in the hands of the supreme-court.

If you are conscientious and responsible, and you act on that responsibility, then all will be well for our society, but if you live and act like a bunch of spineless bugs asleep with complacency, then even a thousand Naskars, Chomskys and Platos would fall short to put the society straight.

13. Gospel of Good Citizenry

What happens when the people do not stand up and act, or when the people do stand up but are crushed by the authoritative power of crooked politicians - what then! That's where civil service comes in. Till we successfully build a meritocratic society, the integrity of a democracy remains predicated on the integrity of the civil servants.

Civil servants are the first defenders of democracy, against crooked politicians as well as angry, mindless mobs.

Some may argue, how can we say that all politicians are crooked, and indeed I agree that all politicians are not crooked, and I'll continue to support such individuals no matter their political affiliations, like I have been supporting Bernie Sanders, Barack Obama, Joe Biden, Kamala Harris and AOC (AOC is a bit rough around the edges, but one day she'll make a great POTUS), but when the concept of politics is founded on popularity and not on training and education, then expecting a capable politician from such a system is the same as expecting to win the lottery - you may win at rare occasions, but losing is way more likely, no matter who you choose. So to change this we have to build a system where it'll be more likely for us to receive a good politician, may not be a great one but good and functional no less, as a leader instead of a crooked one.

However, discrimination and other inhumanities won't magically disappear from the society just because we have a good politician in power - it's only a step in the right direction. To actually eliminate inhumanities from the world all humans, politicians, civil servants and civilians alike, must work hand in hand with accountability.

Let me elaborate. Racism doesn't simply disappear just because we now have a good and proper President in the White House. Look at Canada from example. Justin Trudeau may not be the best PM in the world, but he is not a bigoted one either, yet it hasn't made Canada free from discrimination, misogyny and bigotry - why - because elimination of savagery from our society requires all of us, not just the political authorities. All of us must stand up to all forms inhumanity and say "that's enough - no more".

Having good politicians is important, but more important is to have good citizens.

Who are the citizens - what are the citizens - why are the citizens - the answer to all these questions are the same - it's human. Every human who is a human is also a citizen, and every citizen who is a human has the duty to feel, think and act as a human - and who is to determine how a human is to feel, think and act - nobody - it's not the interest of any one person that is to be placed at the foundation of the determination on how to feel, think and act as a human - that determination is to be made upon the foundation of life - and what is life - life is ascension - life is assimilation - life is unification - any individual that realizes this, can determine quite easily how to feel, think and act as a human.

14. Observation over Cynicism

Decisions of life must be made based on the needs of the living, not of the dead. Let me give you an example. If you live in a different time than mine, then before accepting my words you must contemplate on how much of my work is applicable to the needs of your particular time - after you have done that you can make use of my work accordingly. But mark you, I am not asking of you to be a cynic, for cynicism breeds more confusion and insecurity, not clarity and sanity - what I am asking is for you to be observant. The more advanced we become as a species, the more observant we must become.

Too much comfort is not healthy, for it makes us lazy, and when we are lazy we stop growing.

As a matter of fact, when we are lazy, we stop living. That's why I have said this before and I say it again - do not fall prey to luxury - do not be consumed by luxury, for luxury is the poison in life. Recognize what you need and have that much, not more, and put a little bit of the rest of your available resources in savings and use the remaining to help others - to lift others.

Remember, a human being is the true magic of nature. It is the greatest miracle of all, and also the truest one. The world is filled with stories of supernatural miracles but they haven't made the world any better, because to make this world a better place what's needed is natural miracles, the miracles that have no supernatural nonsense involved, the miracles that have no fiction involved. Fiction may have its place in society, but only as fiction, only as a tool to empower creativity, not as facts.

Fiction is healthy for the mind so long as you do not confuse them with facts.

15. Goodwill Gospel

The only way you can see someone turn water into wine without resorting to trickery is if you are already drunk. That's not the kind of miracle that the world needs - what the world needs is the miracle of kindness - it needs the miracle of acceptance - it needs the miracle of unity in diversity.

There are countless stories of miracles across the planet, especially around religious figures, because they were all made up to engender faith in those figures. Ignorance made people associate divinity with the supernatural. And such ignorance may have suited the people of that time, but we live in a different time now, therefore, it's our responsibility to look at these stories with a fresh perspective and observe the metaphorical lessons behind them if any.

Whether Jesus turning water into wine - whether Mecca moving around following the feet of Nanak - whether Krishna lifting a mountain on his finger - it's all fiction - which were written not to inform the public but to make them accept a certain figure as the authority of their lives. And such behavior may have been acceptable back in those days when ignorance and prejudice were the default mode of thinking, it has no place in today's society.

The point is, we do not need to make up supernatural stories to believe in the goodness of great characters - goodness and greatness do not require magical nonsense to survive in a society of civilized beings. I accept many of these figures as my predecessors in the path of service - they were no magical beings, they were just plain, ordinary, flesh and blood humans who just couldn't stay indifferent to the sufferings of the society - hence they did everything in their power to alleviate them - so do I, and so must every single being who has recognized their responsibility towards society.

Unless a bunch of lionhearts are sleepless for society, the society can never sleep in peace.

I am sleepless - Jesus was sleepless - Gautama was sleepless - Nanak was sleepless - Shankara was sleepless - and so were many - and so must be many more, so that someday humankind may know real peace and harmony. You see - prayers and meditation may bring comfort to you personally, but to make a direct contribution to society, you must not sit or bow in meditation or prayer, rather you must stand up and vow that you won't stop till you actually turn your society into a decent place suitable for human living.

Peace will come chasing after you, once you lose the self in the service of the helpless and the destitute - once you find yourself not in yourself but in others. I looked into the mirror, I couldn't find myself, I looked into you and there I was. This is the foundation of my being - this is the foundation of my work.

Often pastors as well as my fellow scientists ask me - are you a believer or non-believer - I smile and reply, all I care about is to abolish the barriers that people have raised among themselves. I work to create living christs in every field of society to take care of every corner of the world. I am neither a believer, nor a non-believer, the very question is irrelevant to me.

What does the river, the trees, the sun care about the concept of God - their sole purpose is to sustain life on earth - and so is mine. Christ made his sacrifice for his love for society, I have made my sacrifice for my love for society, and I believe strongly more than anything else, that this tradition of sacrifice will continue through many more Christs and Naskars, so long as there is human suffering on earth. Remember, the only absolute truth in the world of humans is love – there is nothing higher, there is nothing greater.

16. When Calls The Future

I don't care about belief or disbelief - I don't care about socialism or capitalism - I don't care about facts or fiction - all these can either exist as servant of humankind or not exist at all. That is why I call people my own, not ideologies. I can take the good from an ideology, but just because there are certain things that are good about an ideology doesn't mean that that ideology is absolutely flawless.

No ideology is flawless, for the very existence of an ideology is a stark expression of sectarianism - of separatism - an ideology exists like a golden wall - it may look appealing and rather magnificent to some, but still it's a wall - a wall that separates people from people.

However, again, let's go very slow - for these are extremely grey areas. Just because I do not approve of ideological loyalty, doesn't mean ideologies will magically disappear from the face of the planet. We like it or not - ideologies are part of the societal fabric - they are part of the societal reality - they are the part of our present. And future is born out of the present not separate from it.

The first step of building a good future is to recognize the reality of the present and then work through that reality. With one hand I support the present and with another I build the future. With one hand I support patriotism and with another I work towards universalism. With one hand I support good democracy and with another I work towards meritocracy. With one hand I support good politicians and with another I build the future without politics.

Remember, envisioning a beautiful future will attain nothing unless you can grasp the reality of the present. Observe the present, empower the good of the present and work through it to build the beautiful future of your vision. Eventually all streams will come together in the name of humanity, and in the meantime, we must embrace the good of all streams while discarding the bad.

17. We Are All Idiots

Inability to recognize reality is not insight, it's blindness. There is no way towards the future except through the present. Present is the gateway to the future. So, whatever vision you hold of the future, if it doesn't involve the present, then it's not a vision but a delusion.

Here we must ask a rather significant question - what is the difference between an illusion and a delusion? And the answer is, anything that is not real but seems real is an illusion and an illusion that is harmful for the individual and the collective is a delusion. Many conspiracy theories can be considered delusion for they have terrible implications in the society, especially those that boast ignorance over well-established scientific conviction.

The point is, just because you don't have expertise in something doesn't mean you are worthless as a person. In fact, admitting ignorance is the first step towards knowledge. Let me give you a few examples to put you at ease in your efforts to admit your own shortfalls.

I am lousy in mathematics, I am terrible in modern physics, I am lazy in sports, and I know nothing about diplomacy. I have spent my life in understanding human nature, outside of it I know nothing about nothing. Now it's your turn. Admit your shortfalls my friend - admit it without shame, because,

we may each be good at certain things but all of us are a bunch of dum-dums in most things.

Admit it, and you'll feel a sense of liberation that no amount of intellectual arrogance can give you. Ignorance is not bliss, but acceptance of ignorance is. After you accept your ignorance, you can start to learn. Remember, love of knowledge gives you understanding whereas arrogance of knowledge gives you blindness. So before you know anything else, know that you know nothing - then proceed from there.

18. Arise, Generation Assimilation

No matter how high you soar, the moment you lose touch with the ground is the moment you become irrelevant to society, to the world, to humankind. And one who is irrelevant to humankind, is anything but human. Remember, we'll either grow together as a collective or perish alone as individuals.

Life of the individual is not of the individual alone, it belongs to the collective. In fact, the entire human society is a living, breathing organism, where each individual acts as a cell - as such, the state of each cell is imposed on the state of the entire societal organism. So your thoughts matter - your emotions matter - your behavior matters - not just for you, but for the entire society.

That is why I say - the society is my life and I'll die protecting it. Society flows through my veins - society flows through my nerves - society flows through my bones - my salvation is in the upliftment of the society - especially of those who are helpless - those who are destitute - those who are discriminated. What's the point of life if it doesn't come to the use of society!

The integrity of the society is predicated on the sacrifice of a handful of lionhearts and on the everyday accountability of the rest of the humans.

You don't need to give up all to come to the use of society - just have a little accountability in your life towards your society - that would be enough - that would be enough to heal the wounds - that would be enough to reduce suffering - that would be enough to instill equality, justice and inclusion. Indifference is inhuman, discrimination is demonic, silence is savagery, prejudice is parochial - the answer to all these is accountability.

This accountability woke up in me in my early teenage years. Of course I had other drives as well, like any healthy teenage kid is supposed to have during puberty, but for some unknown reason as I gained towards my twenties, those drives began to lose their grip and in their place a new invincible drive began to take hold - the drive for social uplift - the drive for humaneness - the drive for humanizing the entire humankind - which eventually led me to what I am today.

Soon my body will perish but my work will continue to create doctors with character and accountability - teachers with character and accountability - politicians with character and accountability - civil servants with character and accountability - scientists with character and accountability - preachers with character and accountability - janitors, bus conductors, waitresses and construction workers with character and accountability - in short, my body will perish but my work will continue to create human beings of character and accountability. When accountability runs through the humans like lifeblood, all will be well for humanity.

You know what the world needs – the world needs thunder-nerves that strike at the sight of injustice and volcanic veins that erupt at the sight of discrimination. And once your accountability is wide awake, your nerves would become thunder-nerves and your veins would become volcanic veins on their own. Then nothing can stop your footsteps from causing a sanctifying tsunami of change in your corner of the world.

Our world has been run by the generation bigotry for way too long, now it's time for the generation assimilation to take charge.

You, me, each one of us - each of us whose veins carry the force of liberty and bravery and whose nerves carry the force of conscience and reason are the embodiment of assimilation. So wake up and rise my generation assimilation and be the bridge to the bridgeless lands - be the voice to the voiceless souls - be the reason to the wild kingdom of bigotry - and above all, be the sapiens to the unsapient society.

BIBLIOGRAPHY

Archer M., (2000), Being Human: The Problem of Agency. Cambridge University Press.

Archer M., (2003), Structure, Agency and the Internal Conversation. Cambridge University Press.

Adolphs R (2003) Cognitive neuroscience of human social behaviour. Nature Rev Neurosci 4: 165–178.

Adolphs R, Tranel D, Damasio AR (2003) Dissociable neural systems for recognizing emotions. Brain Cogn 52: 61–69.

Afton, A. D. (1985). Forced copulation as a reproductive strategy of male lesser scaup: A field test of some predictions. - Behaviour 92, p. 146-167.

Allison T, Puce A, McCarthy G. (2000) Social perception from visual cues: role of the STS region. Trends Cogn Sci 4: 267–278.

Andresen, Jensine, and Robert Forman, eds. Cognitive Models and Spiritual Maps. Bowling Green, Ohio: Imprint Academic, 2000.

Ashbrook, James, and Carol Albright. The Humanizing Brain: Where Religion and Neuroscience Meet. Cleveland, OH: Pilgrim Press, 1997.

Azari, Nina, Janpeter Nickel, Gilbert Wunderlich, Michael Niedeggen, Harald Hefter, Lutz Tellmann, Hans Herzog, Petra Stoerig, Dieter Birnbacher, and Rudiger Seitz. "Neural Correlates of Religious Experience." European Journal of Neuroscience 13, no. 8 (2001)

Agar, N. (2004). Liberal eugenics: In defence of human enhancement. London: Blackwell Publishing.

Alteheld, N., Roessler, G., Vobig, M., & Walter, R. (2004). The retina implant new approach to a visual prosthesis. Biomedizinische Technik, 49(4), 99–103.

Antal, A., Nitsche, M. A., Kincses, T. Z., Kruse, W., Hoffmann, K. P., & Paulus, W. (2004a). Facilitation of visuo-motor learning by transcranial direct current stimulation of the motor and extrastriate visual areas in humans. European Journal of Neuroscience, 19(10), 2888–2892.

Bhat Z, Kumar, S, Bhat H (2015) In vitro meat production. Challenges and benefits over conventional meat production. J Sci Food Agric 14: 241–248

Bernstein R. J., (1967), John Dewey. New York: Washington Square Press.

Bernstein R.J., (1971), Praxis and Action: Contemporary Philosophies of Human Activity. Philadelphia: University of Pennsylvania Press.

Bernstein R.J., (1976), The Restructuring Social and Political Thought.

Bernstein R.J., (1983), Beyond Relativism and Objectivism: Science, Hermeneutics, and Praxis. Philadelphia: University of Pennsylvania Press.

Bernstein R.J., (1986), Philosophical Profiles. Philadelphia: University of Pennsylvania Press.

Bernstein R.J., (1991), New Constellation. Cambridge: MIT Press.

Barash, D. P. (1977). Sociobiology of rape in mallards (Anas platyrhynchos): Responses of the mated male. - Science 197, p. 788-789.

Berger, J. (1986). Wild horses of the great basin: Social competition and population size. - The University of Chicago Press, Chicago.

Birkhead, T. R., Johnson, S. D. & Nettleship, D. N. (1985). Extra-pair matings and mate guarding in the common murre Uria aalge. - Anim. Behav. 33, p. 608-619.

Beauregard, Mario, and Vincent Paquette. "Neural Correlates of a Mystical Experience in Carmelite Nuns." Neuroscience Letters 405, no. 3 (2006)

Benson, Herbert. Timeless Healing: The Power and Biology of Belief. New York: Scribner, 1996

Bogen, J.E.(1995a), 'On the neurophysiology of consciousness: Part I. An overview', Consciousness and Cognition, 4.

Bogen, J.E. (1995b), 'On the neurophysiology of consciousness: Part II. Constraining the semantic problem', Consciousness and Cognition, 4.

Bremner, J. D., R. Soufer, et al. (2001). "Gender differences in cognitive and neural correlates of remembrance of emotional words." Psychopharmacol Bull 35 (3).

Brothers, L. (2002). The social brain: A project for integrating primate behavior and neurophysiology in a new domain. In J. T. Cacioppo et al. (Eds.), Foundations in neuroscience. Cambridge, MA: MIT Press.

Buss, D. D. (2003). Evolutionary Psychology: The New Science of Mind, 2nd ed. New York: Allyn & Bacon.

Buss, D. M. (1989). "Conflict between the sexes: Strategic interference and the evocation of anger and upset." J Pers Soc Psychol 56 (5).

Buss, D. M. (1995). "Psychological sex differences. Origins through sexual selection." Am Psychol 50 (3).

Buss, D. M. (2002). "Review: Human Mate Guarding." Neuro Endocrinol Lett 23 (Suppl 4).

Buss, D. M., and D. P. Schmitt (1993). "Sexual strategies theory: An evolutionary perspective on human mating." Psychol Rev 100 (2).

Blakemore SJ, Decety J (2001) From the perception of action to the understanding of intention. Nature Rev Neurosci 2: 561.

Bruce C, Desimone R, Gross CG (1981) Visual properties of neurons in a polysensory area in superior temporal sulcus of the macaque. J Neurophysiol 46: 369–384.

Buccino G, Vogt S, Ritzl A, Fink GR, Zilles K, Freund HJ, Rizzolatti G (2004) Neural circuits underlying imitation of hand actions: an event related fMRI study. Neuron 42: 323–34.

Colapietro V., (1988), "Human Agency: The Habits of Our Being."

Southern Journal of Philosophy, XXVI, 2, pp. 153-68.

Colapietro V., (1992), "Purpose, Power, and Agency." The Monist, 75, 4 (October) pp. 423-44.

Colapietro V., (2003), "Signs and their vicissitudes: Meanings in excess of consciousness and functionality." Logica, Dialogica, Ideologica, a cure di Susan Petrilli e Patrizia Calefato (Milano: Mimesis), pp. 221-36.

Colapietro V., (2004a), "C. S. Peirce's Reclamation of Teleology." Nature in American Philosophy, ed. Jean De Groot (Washington, D.C.: Catholic University Press of America), pp. 88-108.

Colapietro V., (2004b), "Portrait of a Historicist: An Alternative Reading of Peircean Semiotic." Semiotiche, 2/04 [maggio 2004], pp. 49-68.

Colapietro V., (2006), "Engaged Pluralism: Between Alterity and

Sociality." The Pragmatic Century: Conversations with Richard J. Bernstein (Albany, NY: SUNY Press), pp. 39-68.

Colapietro V., (2009), "Habit, Competence, and Purpose." Forthcoming in The Transactions of the Charles S. Peirce Society. Calder AJ, Keane J, Manes F, Antoun N, Young AW (2000) Impaired recognition and experience of disgust following brain injury. Nature Neurosci 3: 1077–1078.

Carey DP, Perrett DI, Oram MW (1997) Recognizing, understanding and reproducing actions. In: Jeannerod M, Grafman J (eds) Handbook of neuropsychology. Vol. 11: Action and cognition. Elsevier, Amsterdam.

Carr L, Iacoboni M, Dubeau MC, Mazziotta JC, Lenzi GL (2003) Neural mechanisms of empathy in humans: a relay from neural systems for imitation to limbic areas. Proc Natl Acad Sci USA 100: 5497–5502.

Changeux JP, Ricoeur P (1998) La nature et la règle. Odile Jacob, Paris.

Cochin S, Barthelemy C, Roux S, Martineau J (1999) Observation and execution of movement: similarities demonstrated by quantified electroencephalograpy. Eur J Neurosci 11: 1839– 1842.

Chomsky Noam, (2017) Requiem for the American Dream

Chomsky Noam, (2016) Who Rules the World?

Chomsky Noam, (2010) How the World Works

Churchland, P.S. (1986), Neurophilosophy (Cambridge, MA: The MIT Press).

Churchland, P.S. & Ramachandran, V.S. (1993), 'Filling in: Why Dennett is wrong', in Dennett and His Critics:

Demystifying Mind, ed. B. Dahlbom (Oxford: Blackwell Scientific Press).

Churchland, P.S., Ramachandran, V.S. & Sejnowski, T.J. (1994), 'A critique of pure vision', in Large- scale Neuronal Theories of the Brain, ed. C. Koch & J.L. Davis (Cambridge, MA: The MIT Press).

Crick, F. (1994), The Astonishing Hypothesis: The Scientific Search for the Soul (New York: Simon and Schuster).

Crick, F. (1996), 'Visual perception: rivalry and consciousness', Nature, 379.

Crick, F. & Koch, C. (1992), 'The problem of consciousness', Scientific American, 267.

Craig AD (2002) How do you feel? Interoception: the sense of the physiological condition of the body. Nature Rev Neurosci 3: 655–666.

Damasio, A (2003a) Looking for Spinoza. Harcourt Inc. Damasio A (2003b) Feeling of emotion and the self. Ann NY Acad Sci 1001: 253–261.

d'Aquili, Eugene. "Senses of Reality in Science and Religion." Zygon 17, no 4 (1982)

d'Aquili, Eugene. "The Biopsychological Determinants of Religious Ritual Behavior." Zygon 10, no. 1 (1975)

d'Aquili, Eugene. "The Myth-Ritual Complex: A Biogenetic Structural Analysis." Zygon 18, no. 3 (1983)

d'Aquili, Eugene, and Andrew Newberg. The Mystical Mind: Probing the Biology of Religious Experience. Minneapolis: Fortress Press, 1999.

Daly DD. 1958. Ictal affect. Am J Psychiatry.

Damasio, A. (1994) Descartes' Error: Emotion, Reason and the Human Brain. New York, Putnams.

Damasio, A. (1999) The Feeling of What Happens: Body, Emotion and the Making of Consciousness. London, Heinemann.

Darwin, C. (1859) On the Origin of Species by Means of Natural Selection. London, Murray.

Darwin, C. (1871) The Descent of Man and Selection in Relation to Sex. London, John Murray.

Darwin, C. (1872) The Expression of the Emotions in Man and Animals. London, John Murray; also published 1965, Chicago, University of Chicago Press.

Dawkins, M.S. (1987) Minding and mattering. In C. Blakemore and S. Greenfield (eds) Mindwaves. Oxford, Blackwell, 151-60.

Dawkins, R. (1976) The Selfish Gene. Oxford, Oxford University Press; a new edition, with additional material, was published in 1989.

Dawkins, R. (1986) The Blind Watchmaker. London, Longman.

Di Pellegrino G, Fadiga L, Fogassi L, Gallese V, Rizzolatti G (1992) Understanding motor events: A neurophysiological study. Exp Brain Res 91: 176–80.

Deikman, A.J. (2000) A functional approach to mysticism. Journal of Consciousness Studies 7(11-12), 75-91.

Delmonte, M.M. (1987) Personality and meditation. In M. West (ed.) The Psychology of Meditation. Oxford, Clarendon Press, 118-32.

Dennett, D.C. (1987) The Intentional Stance. Cambridge, MA, MIT Press.

Dennett, D.C. (1988) Quining qualia. In A.J. Marcel and E. Bisiach (eds)

Consciousness in Contemporary Science. Oxford, Oxford University Press, 42-77.

Dennett, D.C. (1991) Consciousness Explained. Boston, MA, and London, Little, Brown and Co.

Dennett, D.C. (1995a) Darwin's Dangerous Idea. London, Penguin.

Dennett, D.C. (1995b) The unimagined preposterousness of zombies. Journal of Consciousness Studies 2(4), 322-6.

Dennett, D.C. (1995c) Cog: steps towards consciousness in robots. In T. Metzinger (ed.) Conscious Experience. Thorverton, Devon, Imprint Academic, 471-87.

Dennett, D.C. (1995d) The path not taken. Behavioral and Brain Sciences 18, 252-3; commentary on N. Block, On a confusion about a function of consciousness. Behavioral and Brain Sciences 18, 227.

Dennett, D.C. (1996a) Facing backwards on the problem of consciousness. Journal of Consciousness Studies 3(1), 4-6.

Dennett, D.C. (1996b) Kinds of Minds: Towards an Understanding of Consciousness. London, Weidenfeld & Nicolson.

Dennett, D.C. (1997) An exchange with Daniel Dennett. In J. Searle (ed.) The Mystery of Consciousness. New York, New York Review of Books, 115-19.

Dennett, D.C. (1998) The myth of double transduction. In S.R. Hameroff, A.W. Kaszniak and A. C. Scott (eds) Toward a Science of Consciousness: The Second Tucson Discussions and Debates. Cambridge, MA, MIT Press, 97-107.

Dennett, D.C. (1998b) Brainchildren: Essays on Designing Minds. Cambridge, MA, MIT Press.

Dennett, D.C. (2001) The fantasy of first person science. Debate with D. Chalmers, Northwestern University, Evanston, IL, February 2001.

Dennett, D.C. (2003) Freedom Evolves. New York, Penguin.

Dennett, D.C. and Kinsbourne, M. (1992) Time and the observer: the where and when of consciousness in the brain. Behavioral and Brain Sciences 15, 183-247, including commentaries and authors' responses.

Dewey J., (1911 [1977]), "Epistemological Realism: The Alleged Ubiquity of the Knowledge Relation." Journal of Philosophy, VIII, 20 (September 28, 1911).

Dewhurst, Kenneth, and A. W. Beard. "Sudden Religious Conversions in Temporal Lobe Epilepsy." British Journal of Psychiatry 117 (1970)

Dewhurst K, Beard AW. Sudden religious conversions in temporal lobe epilepsy. 1970 Epilepsy Behav 2003

Devinsky O, Lai G. Spirituality and religion in epilepsy. Epilepsy Behav 2008.

Devinsky, O., Morrell, MJ, Vogt, BA. (1995) 'Contribution of anterior cingulate cortex to behavior', Brain, 118.

Douglas Stone A., Chapter 24, The Indian Comet, in the book Einstein and the Quantum, Princeton University Press, Princeton, New Jersey, 2013.

E. Horvitz, "One Hundred Year Study on Artificial Intelligence: Reflections and Framing," ed: Stanford University, 2014.

Einstein A. (1925). "Quantentheorie des einatomigen idealen Gases". Sitzungsberichte der Preussischen Akademie der Wissenschaften.

Eckhart Meister, Selected Writings

Egidi R., ed. (1999), "Von Wright and 'Dante's Dream': Stages in a Philosophical Pilgrim's Progress", in In Search of a New Humanism: the Philosophy of G.H. von Wright, ed. by R. Egidi, Kluwer, Dordrecht.

Fadiga L, Fogassi L, Pavesi G, Rizzolatti G (1995) Motor facilitation during action observation: a magnetic stimulation study. J Neurophysiol 73: 2608–2611.

Fogassi L, Gallese V, Fadiga L, Rizzolatti G (1998) Neurons responding to the sight of goal directed hand/arm actions in the parietal area PF (7b) of the macaque monkey. Soc Neurosci Abs 24:257.5.

Frith U, Frith CD (2003) Development and neurophysiology of mentalizing. Philos Trans R Soc Lond B Biol Sci 358: 459.

Farah, M.J. (1989), 'The neural basis of mental imagery', Trends in Neurosciences, 10.

Finlay BL, Darlington RB (1995) Linked regularities in the development and evolution of mammalian brains. Science 268.

Freud, S. "The Interpretation of Dreams", 1900

Freud, S. "Selected papers on hysteria and other psychoneuroses" Journal of Nervous and Mental Disease 1909.

Freud, S. "The Origin and Development of Psychoanalysis", 1910

Freud, S. "Psychopathology of everyday life", 1914

Freud, S. "Beyond the Pleasure Principle", 1920

Frith, C.D. & Dolan, R.J. (1997), 'Abnormal beliefs: Delusions and memory', Paper presented at the May,

1997, Harvard Conference on Memory and Belief.

Gay, Volney, ed. Neuroscience and Religion. Plymouth, UK: Lexington Books, 2009.

Gazzaniga, M. S. (1985). The social brain. New York: Basic Books.

Gazzaniga, M.S. (1993), 'Brain mechanisms and conscious experience', Ciba Foundation Symposium, 174.

Geschwind N. "Behavioural changes in temporal lobe epilepsy". Psychol Med. 1979.

Gellhorn, E., Kiely, W.F. "Mystical states of consciousness: neurophysiological and clinical aspects." J Nerv Ment Dis. 1972;154:399-405.

Gilbert SL, Dobyns WB, Lahn BT (2005) Genetic links between brain development and brain evolution. Nat Rev Genet 6.

Gray JA. The Psychology of Fear and Stress. 2nd ed. New York, NY: Cambridge University Press; 1988.

Gloor, P. (1992), 'Amygdala and temporal lobe epilepsy', in The Amygdala: Neurobiological Aspects of Emotion, Memory and Mental Dysfunction, ed J.P. Aggleton (New York: Wiley-Liss).

Greenspan, S. I. and S. G. Shanker (2004). The first idea: How symbols, language, and intelligence evolved from our early primate ancestors to modern humans. Cambridge, MA: Da Capo Press.

Grady, D. (1993), 'The vision thing: Mainly in the brain', Discover, June.

Gallagher HL, Frith CD (2003) Functional imaging of 'theory of mind'. Trends Cogn Sci 7: 77.

Gallese V, Fogassi L, Fadiga L, Rizzolatti G (2002) Action representation and the inferior parietal lobule. In: Prinz W, Hommel B (eds) Attention & Performance XIX. Common mechanisms in perception and action. Oxford University Press, Oxford.

Gallese V, Keysers C, Rizzolatti G (2004) A unifying view of the basis of social cognition. Trends Cogn Sci 8: 396–403.

Gangitano M, Mottaghy FM, Pascual-Leone A (2001) Phase specific modulation of cortical motor output during movement observation. NeuroReport 12: 1489–1492.

Gangitano M, Mottaghy FM, Pascual-Leone A (2004) Modulation of premotor mirror neuron activity during observation of unpredictable grasping movements. Eur J Neurosci 20: 2193– 2202.

Goldman AI, Sripada CS (2004) Simulationist models of face-based emotion recognition. Cognition 94: 193–213.

Grèzes J, Costes N, Decety J (1998) Top-down effect of strategy on the perception of human biological motion: a PET investigation. Cogn Neuropsychol 15: 553–582.

Grèzes J, Armony JL, Rowe J, Passingham RE (2003) Activations related to "mirror" and "canonical" neurones in the human brain: an fMRI study. Neuroimage 18: 928–937.

Gross CG, Rocha-Miranda CE, Bender DB (1972) Visual properties of neurons in the inferotemporal cortex of the macaque. J Neurophysiol 35: 96–111.

Hari R, Forss N, Avikainen S, Kirveskari S, Salenius S, Rizzolatti G (1998) Activation of human primary motor cortex during action observation: a neuromagnetic study.

Proc. Natl Acad Sci USA 95: 15061–15065.

Hardy, G. H. (1940). Ramanujan. Cambridge: Cambridge University Press.

Hall, Daniel, Keith Meador, and Harold Koenig. "Measuring Religiousness in Health Research: Review and Critique." Journal of Religion and Health 47, no. 2 (2008)

Harris, Sam, Jonas Kaplan, Ashley Curiel, Susan Bookheimer, Marco Iacoboni, and Mark Cohen. "The Neural Correlates of Religious and Nonreligious Belief." PLoS One 4, no. 10 (October 1, 2009)

Halgren, E. (1992), 'Emotional neurophysiology of the amygdala within the context of human cognition', in The Amygdala: Neurobiological Aspects of Emotion, Memory and Mental Dysfunction, ed J.P. Aggleton (New York: Wiley-Liss).

Halligan PW, Fink GR, Marshal JC, Vallar G. 2003. Spatial cognition: evidence from visual neglect. Trends Cogn Sci.

Handbook of Emotions, Edited by Michael Lewis, Jeannette M. Haviland-Jones, and Lisa Feldman Barrett, The Guilford Press; 3rd edition (2010).

Haggard, P., Clark, S. and Kalogeras, J. (2002) Voluntary action and conscious awareness, Nature Neuroscience 5, 382-5. Haggard, P., Newman, C. and Magno, E. (1999) On the perceived time of voluntary actions. British Journal of Psychology 90, 291-303.

Hameroff, S.R. and Penrose, R. (1996) Conscious events as orchestrated space-time selections. Journal of Consciousness Studies 3(1), 36-53; also reprinted in J. Shear (ed.) (1997) Explaining Consciousness-The Hard Problem. Cambridge, MA, MIT Press, 177-95.

Hardcastle, V.G. (2000) How to understand theN in NCC. InT. Metzinger (ed.) Neural Correlates of Consciousness. Cambridge, MA, MIT Press, 259-64.

Harding, D.E. (1961) On Having no Head: Zen and the Re-Discovery of the Obvious. London, Buddhist Society.

Hardy, A. (1979) The Spiritual Nature of Man: A Study of Contemporary Religious Experience. Oxford, Clarendon Press.

Hamad, S. (1990) The symbol grounding problem. Physica D 42, 335-46.

Hamad, S. (2001) No easy way out. The Sciences 41(2), 36-42.

Harre, R. and Gillett, G. (1994) The Discursive Mind. Thousand Oaks, CA, Sage.

Haugeland, J. (ed.) (1997) Mind Design II: Philosophy, Psychology, Artificial

Intelligence. Cambridge, MA, MIT Press.

Hauser, M.D. (2000) Wild Minds: What Animals Really Think. New York, Henry Holt and Co.; London, Penguin.

Hearne, K. (1990) The Dream Machine. Northants, Aquarian.

Hebb, D.O. (1949) The Organization of Behavior. New York, Wiley.

Helmholtz, H.L.F. von (1856-67) Treatise on Physiological Optics.

Hess, EH (1975) "The role of pupil size in communication," Scientific American, 233(5), 110–12.

Heyes, C.M. (1998) Theory of mind in nonhuman primates. Behavioral and Brain Sciences 21, 101-48; with commentaries.

Heyes, C.M. and Galef, B.G. (eds) (1996) Social Learning in Animals: The Roots of Culture. San Diego, CA, Academic Press.

Hilgard, E.R. (1986) Divided Consciousness: Multiple Controls in Human Thought and Action. New York, Wiley.

Hocquette JF (2016) Is in vitro meat the solution for the future? Meat Science 120:

167–176

Hodgson, R. (1891) A case of double consciousness. Proceedings of the Society for Psychical Research 7, 221-58.

Hofstadter, D.R. (1979) Code!, Escher, Bach: An Eternal Golden Braid. London, Penguin.

Hofstadter, D.R. and Dennett, D.C. (eds) (1981) The Mind's I: Fantasies and Reflections on Self and Soul. London, Penguin.

Holland, J. (ed.) (2001) Ecstasy: The Complete Guide: A Comprehensive Look at the Risks and Benefits of

MDMA. Rochester, VT, Park Street Press.

Holmes, D.S. (1987) The influence of meditation versus rest on physiological arousal. In M. West (ed.) The Psychology of Meditation. Oxford, Clarendon Press, 81-103.

Holt, J. (1999) Blindsight in debates about qualia. Journal of Consciousness Studies 6(5), 54-71.

Horgan, J. (1994), 'Can science explain consciousness?', Scientific American, 271.

Holloway RL (1996) Evolution of the human brain. In: Lock A, Peters CR (eds) Handbook of human symbolic evolution. Oxford University Press, Oxford

Iacoboni M, Woods RP, Brass M, Bekkering H, Mazziotta JC, Rizzolatti G (1999) Cortical mechanisms of human imitation. Science 286: 2526–2528.

Iacoboni M, Koski LM, Brass M, Bekkering H, Woods RP, Dubeau MC, Mazziotta JC, Rizzolatti G (2001) Reafferent copies of imitated actions in the right superior temporal cortex. Proc Natl Acad Sci USA 98: 13995–13999.

Jeannerod M (1988) The neural and behavioural organization of goal-directed movements. Clarendon Press, Oxford.

Johnson-Frey SH, Maloof FR, Newman-Norlund R, Farrer C, Inati S, Grafton ST (2003) Actions or hand-objects interactions? Human inferior frontal cortex and action observation. Neuron 39: 1053–1058.

Jackson, F. (1982) Epiphenomenal qualia. Philosophical Quarterly 32, 127-36.

James, W. (1890) The Principles of Psychology (2 volumes). London, Macmillan.

James, W. (1902) The Varieties of Religious Experience: A Study in Human Nature. New York and London, Longmans, Green and Co.

Jansen, K. (2001) Ketamine: Dreams and Realities. Sarasota, FL, Multidisciplinary Association for Psychedelic Studies.

Jay, M. (ed.) (1999) Artificial Paradises: A Drugs Reader. London, Penguin.

Jaynes, J. (1976) The Origin of Consciousness in the Breakdown of the Bicameral Mind. New York, Houghton Mifflin.

Johnson, M.K. and Raye, C.L. (1981) Reality monitoring. Psychological Review 88, 67-85.

Kadim I, Mahgoub O, Baqir S et al. (2015) Cultured meat from muscle stem cells: a review of challenges and prospects. J Integr Agr 14: 222–233

Koski L, Iacoboni M, Dubeau MC, Woods RP, Mazziotta JC (2003) Modulation of cortical activity during different imitative behaviors. J Neurophysiol 89: 460–471.

Krolak-Salmon P, Henaff MA, Isnard J, Tallon-Baudry C, Guenot M, Vighetto A, Bertrand O, Mauguiere F (2003) An attention modulated response to disgust in human ventral anterior insula. Ann Neurol 53: 446–453.

Kandel, E. R. In Search of Memory: The Emergence of a New Science of Mind, W. W. Norton & Company (2007).

Kandel E. R. Schwartz JH, Jessel TM. Principles of neural sciences. New York; McGraw Hill, 2000.

Kanizsa, G. (1979), Organization In Vision (New York: Praeger).

Kaloupek DG, Scott JR, Khatami V. Assessment of coping strategies associated with syncope in blood donors. J Psychosom Res. 1985;29:207-214.

Kanwisher, N. (2001) Neural events and perceptual awareness. Cognition 79, 89-113; also reprinted inS. Dehaene (ed.) The Cognitive Neuroscience of Consciousness. Cambridge, MA, MIT Press, 89-113.

Kapleau, Roshi P. (1980) The Three Pillars of Zen: Teaching, Practice, and Enlightenment (revised edn). New York, Doubleday.

Karn, K. and Hayhoe, M. (2000) Memory representations guide targeting eye movements in a natural task. Visual Cognition 7, 673-703.

Kasamatsu, A. and Hirai, T. (1966) An electroencephalographic study on the Zen meditation (zazen). Folia Psychiatrica et Neurologica Japonica 20, 315-36.

Kaiserman-Abramof, I. R., Graybiel, A. M., & Nauta, W. J. (1980). The thalamic projection to cortical area 17 in a congenitally anophthalmic mouse strain. Neuroscience, 5, 41–52.

Kanold, P. O., Kara, P., Reid, R. C., & Shatz, C. J. (2003). Role of subplate neurons in functional maturation of visual cortical columns. Science, 301, 521–525.

Kennedy, H., & Dehay, C. (1988). Functional implications of the anatomical organization of the callosal projections of visual areas V1 and V2 in the macaque monkey. Behav. Brain Res., 29, 225–236.

Kentridge, R.W. and Heywood, C.A. (1999) The status of blindsight. Journal of Consciousness Studies 6(5), 3-11.

Kihlstrom, J.F. (1996) Perception without awareness of what is perceived, learning without awareness of what is learned. In M. Velmans (ed.) The Science of Consciousness. London, Routledge, 23-46.

Kollerstrom, N. (1999) The path of Halley's comet, and Newton's late apprehension of the law of gravity. Annals of Science 56, 331-56.

Kosslyn, S.M. (1980) Image and Mind. Cambridge, MA, Harvard University Press.

Kosslyn, S.M. (1988) Aspects of a cognitive neuroscience of mental imagery. Science 240, 1621-6.

Kinsbourne, M. (1995), 'The intralaminar thalamic nucleii', Consciousness and Cognition, 4.

Kjaer, Troels, Camilla Bertelsen, Paola Piccini, David Brooks, Jorgen Alving, and Hans Lou. "Increased Dopamine Tone during Meditation- Induced Change of Consciousness." Cognitive Brain Research 13, no. 2 (April 2002)

Kölmel HW. 1985. Complex visual hallucinations in the hemianopic field. J Neurol Neurosurg Psychiatry.

Koenig, Harold. "Research on Religion, Spirituality, and Mental Health: A Review." Canadian Journal of Psychiatry 54, no. 5 (May 2009)

Koenig, Harold, ed. Handbook of Religion and Mental Health. San Diego, CA: Academic Press, 1998

Kraepelin E. Psychiatry: A Textbook for Students and Physicians. New York, NY: Science History Publications; 1990.

Lauglin, Charles, John McManus, and Eugene d'Aquili. Brain, Symbol, and Experience. 2nd ed. New York: Columbia University Press, 1992

Lakoff, G. and M. Johnson (1999). Philosophy in the flesh. Basic Books: New York.

LeDoux, J. E. (1996). The emotional brain. New York: Simon & Schuster.

LeDoux, J.E. (1992), 'Emotion and the amygdala', in The Amygdala:

Neurobiological Aspects of Emo- tion, Memory and Mental Dysfunction, ed J.P. Aggleton (New York: Wiley-Liss).

Levin, D.T. and Simons, D.J. (1997) Failure to detect changes to attended objects in motion pictures. Psychonomic Bulletin and Review 4, 501-6.

Levine,J. (1983) Materialism and qualia: the explanatory gap. Pacific Philosophical Quarterly 64, 354-61.

Levine,J. (2001) Purple Haze: The Puzzle of Consciousness. New York, Oxford University Press. Levine, S. (1979) A Gradual Awakening. New York, Doubleday.

Levinson, B.W. (1965) States of awareness during general anaesthesia. British Journal of Anaesthesia 37, 544-6.

Lewicki, P., Czyzewska, M. and Hoffman, H. (1987) Unconscious acquisition of complex procedural knowledge. Journal of Experimental Psychology: Learning, Memory and Cognition 13, 523-30.

Lewicki, P., Hill, T. and Bizot, E. (1988) Acquisition of procedural knowledge about a pattern of stimuli that cannot be articulated. Cognitive Psychology 20, 24-37.

Lewicki, P., Hill, T. and Czyzewska, M. (1992) Nonconscious acquisition of information. American Psychologist 47, 796-801.

Manthey S, Schubotz RI, von Cramon DY (2003). Premotor cortex in observing erroneous action: an fMRI study. Brain Res Cogn Brain Res 15: 296–307.

Mesulam MM, Mufson EJ (1982) Insula of the old world monkey. III: Efferent cortical output and comments on function. J Comp Neurol 212: 38–52.

Naskar, Abhijit. "Homo: A Brief History of Consciousness", 2015

Naskar, Abhijit. "What is Mind?", 2016

Naskar, Abhijit. "Love, God & Neurons: Memoir of A Scientist who found himself by getting lost", 2016

Naskar, Abhijit. "Principia Humanitas", 2017

Naskar, Abhijit. "We Are All Black: A Treatise on Racism", 2017

Naskar, Abhijit. "Either Civilized or Phobic: A Treatise on Homosexuality", 2017

Naskar, Abhijit. "I Am The Thread: My Mission", 2017

Naskar, Abhijit. "The Bengal Tigress: A Treatise on Gender Equality", 2017

Naskar, Abhijit. "Morality Absolute", 2017

Naskar, Abhijit. "Build Bridges not Walls: In the name of Americana", 2018

Naskar, Abhijit. "Fabric of Humanity", 2018

Naskar, Abhijit. "Lives To Serve Before I Sleep", 2019

Naskar, Abhijit. "Citizens of Peace: Beyond the Savagery of Sovereignty", 2019

Naskar, Abhijit. "The Constitution of The United Peoples of Earth", 2019

Naskar, Abhijit. "Neurons Giveth, Neurons Taketh Away | Abhijit Naskar | TEDxIIMRanchi", 2019 https://www.youtube.com/watch?v=BNX-Q0ySm80

Naskar, Abhijit. "Mission Reality", 2019

Naskar, Abhijit. "Operation Justice: To Make A Society That Needs No Law", 2019

Naskar, Abhijit. "Every Generation Needs Caretakers: The Gospel of Patriotism", 2020

Naskar, Abhijit. "Revolution Indomable", 2020

Naskar, Abhijit. "Servitude is Sanctitude", 2020

Newberg, Andrew, and Jeremy Iversen. "The Neural Basis of the Complex Mental Task of Meditation: Neurotransmitter and Neurochemical Considerations." Medical Hypotheses 61, no. 2 (2003).

Newberg, Andrew. "How God Changes Your Brain: An Introduction to Jewish Neurotheology", CCAR Journal: The Reform Jewish Quarterly, Winter 2016.

Newberg, Andrew, and Stephanie Newberg. "A Neuropsychological Perspective on Spiritual Development." In Handbook of Spiritual Development in Childhood and Adolescence, edited by Eugene Roehlkepartain, Pamela King, Linda

Wagener, and Peter Benson. London: Sage Publications, Inc., 2005

Newberg, Andrew. "The Neurotheology Link An Intersection Between Spirituality and Health", Alternative and Complimentary Therapies, Vol 21 No 1, February 2015.

Newberg, Andrew, Nancy Wintering, Dharma Khalsa, Hannah Roggenkamp, and Mark Waldman. "Meditation Effects on Cognitive Function and Cerebral Blood Flow in Subjects with Memory Loss: A Preliminary Study." Journal of Alzheimer's Disease 20, no. 2 (2010)

Nash, M. (1995), 'Glimpses of the mind', Time.

Nesse RM. Proximate and evolutionary studies of anxiety, stress and depression: synergy at the interface. Neurosci Biobehav Rev. 1999;23:895-903.

Nicolelis, Miguel. (2011) "Beyond Boundaries: The New Neuroscience of Connecting Brains with Machines---and How It Will Change Our Lives", Times Books

O'Hara, K. and Scutt, T. (1996) There is no hard problem of consciousness. Journal of Consciousness Studies 3(4), 290-302, reprinted in J. Shear (ed.) (1997) Explaining Consciousness. Cambridge, MA, MIT Press, 69-82.

O'Regan, J.K. (1992) Solving the "real" mysteries of visual perception: the world as an outside memory. Canadian Journal of Psychology 46, 461-88.

O'Regan, J.K. and Noe, A. (2001) A sensorimotor account of vision and visual consciousness. Behavioral and Brain Sciences 24(5), 883-917.

O'Regan, J.K., Rensink, R.A. and Clark,J.J. (1999) Change-blindness as a result of "mudsplashes." Nature 398, 34.

Ornstein, R.E. (1977) The Psychology of Consciousness (2nd edn). New York, Harcourt.

Ornstein, R.E. (1986) The Psychology of Consciousness (3rd edn). New York, Pehguin.

Ornstein, R.E. (1992) The Evolution of Consciousness. New York, Touchstone.

Penfield W, Faulk ME (1955) The insula: further observations on its function. Brain 78: 445– 470.

Penrose, R. (1994), Shadows of the Mind (Oxford: Oxford University Press).

Penrose, R. (1989), The Emperor's New Mind: Concerning Computers, Minds and The Laws of Physics (Oxford: Oxford University Press).

Persinger, "'I would kill in God's name' role of sex, weekly church attendance, report of a religious experience and limbic lability" Perceptual and Motor Skills 1997.

Persinger "Experimental simulation of the God experience" Neurotheology 2003.

Persinger, M. A. (1993b). Personality changes following brain injury as a grief response to the loss of sense of self: Phenomenological themes as indices of local lability and neurocognitive restructuring as psycho- therapy. Psychological Reports, 72

Persinger, Corradini, Clement, Keaney, et al "Neurotheology and its convergence with neuroquantology" NeuroQuantology 2010.

Persinger, Koren and St-Pierre "The electromagnetic induction of mystical and altered states within the laboratory" Journal of Consciousness Exploration and Research 2010.

Persinger "Case report: A prototypical spontaneous 'sensed presence' of a sentient being and concomitant electroencephalographic activity in the clinical laboratory" Neurocase 2008.

Persinger and Saroka "Potential production of Hughlings Jackson's "parasitic consciousness" by physiologically-patterned weak transcerebral magnetic fields: QEEG and source localization" Epilepsy & Behavior 28 (2013).

Persinger. "The neuropsychiatry of paranormal experiences". J Neuropsychiatry Clin Neurosci 2001.

Persinger. "Neuropsychological bases of god beliefs", New York: Praeger, 1987

Persinger. "Temporal lobe epileptic signs and correlative behaviors displayed by normal populations", Journal of General Psychology, 1986

Perry BD, Pollard R. Homeostasis, stress, trauma, and adaptation. A neurodevelopmental view of childhood trauma. Child Adolesc Psychiatr Clin N Am. 1998;7:33.

Paré, D. & Llinás, R. (1995), 'Conscious and preconscious processes as seen from the standpoint of sleep-waking cycle neurophysiology', Neuropsychologia, 33.

P. S. de Laplace. Essai Philosophique sur les Probabilites [1814], in Academy des Sciences, Oeuvres Complotes de Laplace, Vol. 7, Gauthier-Villars, Paris (1886).

Perrett DI, Harries MH, Bevan R, Thomas S, Benson PJ, Mistlin AJ, Chitty AJ, Hietanen JK, Ortega JE (1989) Frameworks of analysis for the neural representation of animate objects and actions. J Exp Bio 146: 87–113.

Phillips ML, Young AW, Senior C, Brammer M, Andrew C, Calder AJ, Bullmore ET, Perrett DI, Rowland D, Williams SC, Gray JA, David AS (1997) A specific neural substrate for perceiving facial expressions of disgust. Nature 389: 495–498.

Phillips ML, Young AW, Scott SK, Calder AJ, Andrew C, Giampietro V, Williams SC, Bullmore ET, Brammer M, Gray JA (1998) Neural responses to facial and vocal expressions of fear and disgust. Proc R Soc Lond B Biol Sci 265: 1809–1817.

Puce A, Perrett D (2003) Electrophysiological and brain imaging of biological motion. Philosoph Trans Royal Soc Lond, Series B, 358: 435–445.

Ramachandran VS. Behavioral and magnetoencephalographic correlates of plasticity in the adult human brain. Proc Natl Acad Sci USA 1993; 90: 10413–20.

Ramachandran VS. Phantom limbs, neglect syndromes, repressed memories, and Freudian psychology. Int Rev Neurobiol 1994; 37: 291–333.

Ramachandran VS. Plasticity and functional recovery in neurology. Clin Med 2005; 5: 368–73.

Ramachandran VS, Hirstein W. The perception of phantom limbs. The D. O. Hebb lecture. Brain 1998; 121: 1603–30.

Ramachandran VS, Rogers-Ramachandran D, Cobb S. Touching the phantom limb. Nature 1995; 377: 489–90.

Ramachandran VS, Rogers-Ramachandran D. Phantom limbs and neural plasticity. Arch Neurol 2000; 57: 317–20.

Ramachandran VS, Rogers-Ramachandran D. It's all done with mirrors. Sci Am Mind 2007; 18: 16–9.

Ramachandran VS, Rogers-Ramachandran D. Sensations referred to a patient's phantom arm from another subjects intact arm: perceptual correlates of mirror neurons. Med Hypotheses 2008; 70: 1233–4.

Ramachandran VS, Rogers-Ramachandran D, Stewart M. Perceptual correlates of massive cortical reorganization. Science 1992; 258: 1159–60.

Rizzolatti G, Craighero L (2004) The mirror-neuron system. Annu Rev Neurosci 27: 169–192.

Rizzolatti G, Fogassi L, Gallese V (2001) Neurophysiological mechanisms underlying the understanding and imitation of action. Nature Rev Neurosci 2:661–670.

Rock I, Victor J. Vision and touch: an experimentally created conflict between the two senses. Science 1964; 143: 594–6.

Rose´n B, Lundborg G. Training with a mirror in rehabilitation of the hand. Scand J Plast Reconstr Surg Hand Surg 2005; 39: 104–8.

Royet JP, Plailly J, Delon-Martin C, Kareken DA, Segebarth C (2003) fMRI of emotional responses to odors: influence of hedonic valence and judgment, handedness, and gender. Neuroimage 20: 713–728.

Rozin R Haidt J and McCauley CR (2000) Disgust. In: Lewis M, Haviland-Jones JM (eds) Handbook of Emotion. 2nd Edition. Guilford Press, New York, pp 637–653.

Saxe R, Carey S, Kanwisher N (2004) Understanding other minds: linking developmental psychology and functional neuroimaging. Annu Rev Psychol 55: 87–124.

S. J. Russell and P. Norvig, Artificial intelligence: a modern approach (3rd edition): Prentice Hall, 2009.

Schienle A, Stark R, Walter B, Blecker C, Ott U, Kirsch P, Sammer G, Vaitl D (2002) The insula is not specifically involved in disgust processing: an fMRI study. Neuroreport 13: 2023–2026.

Showers MJC, Lauer EW (1961) Somatovisceral motor patterns in the insula. J Comp Neurol 117: 107–115.

Singer T, Seymour B, O'Doherty J, Kaube H, Dolan RJ, Frith CD (2004) Empathy for pain involves the affective but not the sensory components of pain. Science 303: 1157–1162.

Smith A (1759) The theory of moral sentiments (ed. 1976). Clarendon Press, Oxford.

S. N. Bose (1924). "Plancks Gesetz und Lichtquantenhypothese". Zeitschrift für Physik. 26 (1): 178–181.

Sprengelmeyer R, Rausch M, Eysel UT, Przuntek H (1998) Neural structures associated with recognition of facial expressions of basic emotions Proc R Soc Lond B Biol Sci 265: 1927–1931.

Strafella AP, Paus T (2000) Modulation of cortical excitability during action observation: a transcranial magnetic stimulation study. NeuroReport 11: 2289–2292.

Simonsen R (2015) Eating for the future: veganism and the challenge of in vitro meat. In: Stapleton P, Byers A (Hg). Biopolitics and utopia. Palgrave Macmillan, New York (2015), S 167–190

Tanaka K (1996) Inferotemporal cortex and object vision. Ann Rev Neurosci. 19: 109–140.

Tesla N. "My Inventions", 1919

T. R. Society, "Machine learning: the power and promise of computers that learn by example," ed. The Royal Society, 2017.

Tomasello M, Call J (1997) Primate cognition. Oxford University Press, Oxford.

Tremblay C, Robert M, Pascual-Leone A, Lepore F, Nguyen DK, Carmant L, Bouthillier A, Theoret H (2004) Action observation and execution: intracranial recordings in a human subject. Neurology. 63: 937–938.

Umilta MA, Kohler E, Gallese V, Fogassi L, Fadiga L, Keysers C, Rizzolatti G (2001) "I know what you are doing": a neurophysiological study. Neuron 32: 91–101.

Von Wright G.H., (1963), Norm and Action. A Logical Inquiry, Routledge & Kegan Paul, London.

Von Wright G.H., (1976), "Determinism and the Study of Man", in Essays on Explanation and Understanding, ed. by J. Manninen and R. Tuomela, Reidel, Dordrecht.

Von Wright G.H., (1977), "What is Humanism?", The Lindlay Lecture, University of Arkansas, Lawrence, Kansas.

Von Wright G.H., (1979), "Humanism and the Humanities", in Philosophy and Grammar, ed. by S. Kanger and S. Öhman, Reidel, Dordrecht, pp. 1-16. Reprinted in von Wright (1993).

Von Wright G.H., (1980), Freedom and Determination, North-Holland Publishing Co., Amsterdam.

Von Wright G.H., (1985), Of Human Freedom, The Tanner Lectures on Human Values,

Vol. VI, ed. by S. M. McMurrin, University of Utah Press, Salt Lake City, pp. 107-70. Reprinted in von Wright (1998).

Von Wright G.H., (1993), The Tree of Knowledge and Other Essays, Brill, Leiden.

Von Wright G.H., (1997), "Progress: Fact and Fiction", in The Idea of Progress, ed. by A. Burgen et al., W. de Gruyter, Berlin, pp. 1-18.

Von Wright G.H., (1998), In the Shadow of Descartes: Essays in the Philosophy of Mind, Kluwer, Dordrecht.

GOOD SCIENTIST

GOOD SCIENTIST

GOOD SCIENTIST

www.ingramcontent.com/pod-product-compliance
Lightning Source LLC
Chambersburg PA
CBHW031630210526
45464CB00004B/1826